Pedro Augusto Izidoro Pereira
Rochele Amorim Ribeiro

The inclusion of BIM in the Civil Engineering course

Pedro Augusto Izidoro Pereira
Rochele Amorim Ribeiro

The inclusion of BIM in the Civil Engineering course

Study of a Pedagogical Plan

ScienciaScripts

Imprint

Cover image: www.ingimage.com

This book is a translation from the original published under ISBN 978-613-9-67970-6.

Publisher:
Sciencia Scripts
is a trademark of
Dodo Books Indian Ocean Ltd. and OmniScriptum S.R.L publishing group

120 High Road, East Finchley, London, N2 9ED, United Kingdom
Str. Armeneasca 28/1, office 1, Chisinau MD-2012, Republic of Moldova, Europe
Printed at: see last page
ISBN: 978-620-8-18961-7

SUMMARY

1 SUMMARY OF THE INITIAL PLAN

The aim of this study was to identify pedagogical strategies for approaching BIM in undergraduate Civil Engineering teaching, taking into account CES/CNE Resolution 11 (MEC, 2002) by better integrating knowledge of the application of the BIM tool (Building Information Modeling) with the needs of teaching Design and Project in Civil Engineering in all existing interfaces.

This research was proposed and developed during the current year 2013/2014, investigating the current conditions of university training for Civil Engineering students with regard to training in the use of the BIM platform in the national scenario, through bibliographical research and evaluation of questionnaires answered by professors from Brazilian universities, both private and public.

In order to accomplish what was proposed in this work, in addition to collecting data through questionnaires, a bibliographical review was first carried out on BIM technology and its application in both the educational and professional spheres, in order to gain a better grounding in the subject.

This report contains the motivation behind the project, a description of the methodology used and an analysis of the data collected to obtain the results.

1.1 WORK SCHEDULE

The activities carried out for this work are listed in Table 1.

Table 1 - Work schedule.

	2013					2014						
	AUG	SET	OUT	NOV	TEN	JAN	FEB	MAR	APR	MAY	JUN	JUL
Bibliographical review	■	■	■	■								
Preparation and application of questionnaires;				■	■	■	■	■				
Discussion of Results									■	■		
Conclusion, Preparation of Final Report and Preparation of Articles											■	■

2 INTRODUCTION

Successful development of building and construction projects depends above all on the ability to make different information about the elements involved in the process compatible. The development of a construction project involves a variety of factors, such as information on material specifications, labor costs, execution time, compatibility of infrastructure installations, among others, which need to be handled simultaneously in the planning and execution process. In this way, the civil engineering professional must integrate all this information in a systematic and accurate way in order to make the best planning decisions and comply with the proposed construction schedule (HABITARE, 2003a; 2003b).

Civil Engineering therefore uses computational tools to optimize the process of preparing building projects, with Computer-Aided Design (CAD) being the main exponent. However, due to the increasing complexity of construction projects, it is essential to aggregate information in alphanumeric data (costs, quantity of material, quality of material) in the spatial entity, enabling integrated management of this information. One solution to this problem is to make use of Building Information Modeling, known as BIM.

Unlike the 3D CAD model, which only represents the three-dimensional features of the building elements, the BIM model allows the association of information about the components that make up these elements. In other words, in the BIM model, building elements are linked to graphical, three-dimensional, quantitative and parametric attributes, allowing for the generation of descriptive construction documents, such as two-dimensional representations (e.g. floor plans, sections, façades, details), material performance analyses (e.g. acoustics, thermal comfort), budget spreadsheets and physical-financial schedules. In addition, using the BIM platform, it is possible to develop a collaborative project, which allows the intervention of several teams responsible for the project, considering the updates of the project in a consistent and non-redundant way (SMITH and TARDIV, 2009; CEROVSEK, 2011; EASTMAN et. al., 2011).

The training of Civil Engineering professionals in the use of the BIM platform is a growing concern in undergraduate courses and in university research, which is more evident on the international scene, but is growing on the national scene. As a result, there is a need for a pedagogical strategy that can better integrate knowledge of the application of the BIM tool with the needs of teaching Design and Project in Civil Engineering in all areas: architectural, structural, infrastructure, among others.

Therefore, this study investigated the current conditions of undergraduate Civil Engineering education in terms of training in the use of the BIM platform in Brazil. Through bibliographical research and the evaluation of questionnaires answered by professors at Brazilian universities, the main objective of this research was to identify pedagogical strategies for approaching the BIM platform in undergraduate Civil Engineering teaching. The secondary objectives were: (i) to survey scientific research, especially at postgraduate level, that addresses the issue of BIM in undergraduate teaching; (ii) to survey the profile of the teaching staff of undergraduate courses that work with BIM, with emphasis on the national scenario; (iii) to survey the possibilities for the integrated inclusion of BIM in the disciplines of the undergraduate Civil Engineering course at UFSCar, taking into account CES/CNE Resolution 11 (MEC, 2002). It is hoped that the results of this research will help define pedagogical strategies to enable the inclusion of BIM in undergraduate Civil Engineering teaching at UFSCar, through an integrated and multidisciplinary approach to BIM in the course curriculum.

3 BUILDING INFORMATION MODELING - BIM

The dynamics of this approach to BIM are as follows: first, there is a historical survey of the technology; then there is a definition of BIM, its advantages and its applications, followed by a classification of the stages of its use in the market and in teaching. Subsequently, the use of Building Information Modeling is elucidated, both in the job market and in teaching, considering national and international experience, ending with existing strategies for teaching the methodology.

3.1 BRIEF HISTORY

CAD tools, fundamental to the construction process as a whole, have undergone a major evolution over time, and this change has been potentially significant in the AEC (Architecture, Engineering and Construction) industry.

The first CAD technologies were simple electronic drawing boards that created drawings in two dimensions (2D). In the late 1970s, 3D CAD was developed using three-dimensional models, but they were expensive and overloaded the capacity of computers at the time. This improvement stemmed from the real need to develop technologies based on three-dimensional models created specifically to meet the needs of the construction industry. In this respect, the use of BIM has proved to be effective, since it manages to bring together all the information from the life cycle of a building. (PIMENTEL, 2012)

In 1987, the first software with BIM tools was launched in Hungary: Archicad, by Graphisoft. At the same time, the first versions of Autodesk's AutoCAD and Bentley Microstation were launched - both did not have BIM technology, but were the most widespread products at the time.

With technological advances, in 2002 Autodesk bought the parametric model developed by Revit Technology Corporation. Because it was innovative, the product spurred other major system developers to create their own products. Bentley launched Bentley Building Information Modelling, Nemetschek launched AllPlan 2003 and Graphisoft launched the ninth version of ArchiCad. From then on, this software became known as Parametric Virtual Building Modeling, or simply Building Information Modeling (BIM) (TSE et al., 2005).

The current trend in the AEC industry is the integrated solution, better known as 5D CAD: BIM plus time and cost information, optimizing the design process and minimizing construction errors. In other words, BIM can contain information about the building, such as aspects of construction, management, maintenance and operation. Figure 1 illustrates the evolution of CAD systems through a timeline.

Figure 1 - Evolution of CAD technology

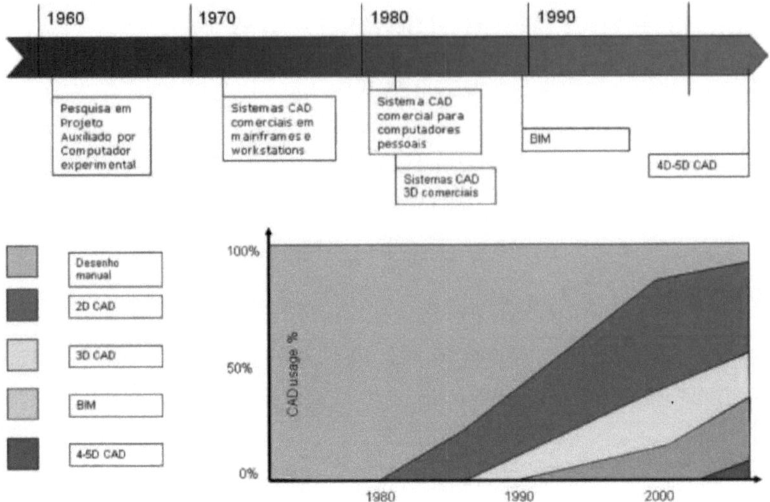

SOURCE: PIMENTEL, 2012.

3.2 CONCEPTUALIZATION OF BIM

The term BIM, *Building Information Modeling*, goes beyond the meaning behind this acronym (RUSCHEL et al., 2013). For PENTILLÂ (2006), BIM "[...] is a methodology for managing the essence of the project and construction or development data in digital format throughout the building's life cycle [...]". In other words, it is a set of information created and maintained throughout the life cycle of a building. SUCCAR (2009) adds that this methodology arises from an interrelated set of policies, processes and technology.

For PIMENTEL (2012) *apud* PEACOCK (2003), BIM is an acronym that encompasses the entire creative data process from the design to the construction of a building. According to PEACOCK, this becomes more evident when the terms are separated: *Building*, which refers to the installation or building being studied; *Information*, which relates to the data obtained from it; and *Modeling*, which relates to the model that controls the information for better management of the installation or building.

BIM technology not only encompasses the benefits of 3D CAD, but also manages to bring together all the components of the project, both qualitative and quantitative, such as the durability, resistance and useful life of the building. This allows for a better understanding and visualization of the project as a whole, as well as greater integration between designers. Construction incompatibilities can also be detected well in advance, in addition to automatically generating material quantities and data on costs and execution times.

This three-dimensional geometric model facilitates or allows simulations of almost all the aspects needed to define a project, such as the structural system, the hydrosanitary system, estimates and budgets.

BIM is a methodology that digitally explores the essential characteristics of a building during its design phase, reducing errors during the construction process, planning and useful life of the building.

The first benefits of adopting Building Information Modeling-based design are: visualization, collaboration and reduction. Visualization is directly related to the 3D model, and is beneficial as it increases understanding of the project. Collaboration refers to the cooperative action of the various members of the project team that is facilitated in the BIM process. And reduction refers to the reduction of conflicts between projects, waste and risk (KYMMELL, 2008).

The design process based on BIM technology leads to significant changes. According to EASTMAN *et al* (2011), the first and most notable is the improvement of processes at each stage of design and construction, reducing the number and severity of problems associated with drawings. It also promotes early collaboration between engineer, architect, contractor and others, which requires greater knowledge from the outset. In short, BIM-based design brings about changes related to (a) team collaboration; (b) legal aspects for documenting ownership and production; (c) the practice and use of information and (d) the implementation of new organizational processes.

For FU et al. (2006), the use of BIM technology contributes to the standardization of design and planning information, which is commonly considered fragmented in all aspects of building design.

TSE et al. (2005) carried out a survey in Hong Kong and found that the biggest motivation for using BIM was the possibility of creating views and schedules dynamically and automatically. Two other major reasons were: instantaneous changes to all drawings and schedules, and the creation of a single project file.

Currently, more and more incentives are being created for the use of BIM. Large companies such as Petrobras in Brazil and public bodies such as London City Hall are demanding the use of the BIM platform in their projects. In addition, construction companies are already helping to better integrate BIM into the market by creating catalogs of their products in BIM.

As with the whole construction process, the understanding of what BIM technology is is still small compared to its scope. There are approaches that are restricted to computer programs, and those that are more in-depth and appropriate, consisting of a collaborative design process based on an integrated model of building information (RUSCHEL *et al.*, 2011).

Some barriers to the use of BIM, according to TSE et al. (2005), were: a complex and time-consuming modeling process, a lack of staff training and technical support, high costs for acquiring extra files, and the unavailability of a free version of the software for testing.

3.3 CLASSIFICATION OF THE STAGES IN THE USE OF BIM

The inclusion of the BIM platform in the preparation and documentation of engineering and architectural projects is a growing process in the global context, although different stages of adoption of the platform can be observed. According to PORWAL and HEWAGE (2013), four levels of maturity have been identified in the use of computer tools for project development: (1) Level zero, characterized by the use of CAD2D as tools for two-dimensional entities, such as lines, arcs, texts, among others; (2) Level one, characterized by the use of 2D and 3D CAD simultaneously, with the elaboration of three-dimensional models; (3) Level two, such as the use of BIM, which allows for the parameterization of objects and the development of collaborative projects; (4) Level three, characterized by the integration and interoperability of data via the Internet, with an emphasis on the management of the construction life cycle.

While in the international scenario there is research and project development at levels two and three (SUCCAR, 2009; SACKS, 2010; LEE, 2012; ZHANG, 2013), in the Brazilian scenario, despite the fact that there is already research at levels one and two (COELHO and NOVAES, 2008; MONTEIRO, 2011; ANDRADE, 2012; SANTOS and BARISON, 2013), the maturity of the use of computational tools in Civil Engineering projects is mostly at level zero.

Finally, once BIM has been implemented into the working method, many authors point out that complete adoption does not occur immediately, but rather over a sequence of stages, until it is fully understood (TOBIN, 2008) (SUCCAR, 2009) (JERNIGAN, 2007). According to RUSCHEL et al., 2011, TOBIN, 2008 and SUCCAR, 2009. The stages of BIM implementation are classified into three:

INTERNSHIP 1. With a focus on parametric modeling, technology is only used as a specific tool in the graphic representation disciplines, as the work process is still individualized, without the involvement and collaboration of other disciplines in the course. It is generally restricted to a specific phase of the process (design, construction or operation).

STAGE 2. With a focus on collaboration, the technology has a multidisciplinary nature between one or two phases of the design process, involving up to two different themes, such as architecture and structure, or cost management. The development of the work relies on the interoperability of project information, making it possible to manipulate a single three-dimensional model between all the teams involved in the work. The process is also interactive and still asynchronous, but with improved interoperability between the teams involved. At this stage, time (4D) and cost (5D) information is already added to the model.

STAGE 3: Focusing on the shared and collaborative creation of the building model synchronously through network integration, the development of the work includes complex analyses already in the initial stages of conception, such as the analysis of design conflicts, since the possibilities for interoperability are expanded through open protocols and virtual work environments, so that all the agents involved in the project can contribute collectively within the specificities of their disciplines. Here, the models are already built in a centralized database on which all participants can interact (interdisciplinary view of the building).

According to CHECCUCCI (2014), the adoption of BIM requires team training, the acquisition of software and hardware, as well as the development of new technologies and techniques based on the collaborative and integrative process. For the author, prior planning is of the utmost importance when implementing this technology, seeking to understand the objectives for its use, the need and capacity to invest in it, the time to be dedicated to the transition from one technology to another, among others.

3.4 BIM IN THE JOB MARKET

The BIM concept gained momentum at the end of the 1970s, due to major economic changes resulting from the globalization of markets and increased pressure on companies to improve processes in order to reach a market that was increasingly demanding in terms of deadlines, quality and costs (SOUZA *et al.*, 2009).

Various works on product modeling in the AEC industry were developed in the 1970s and 1980s in the United States and Europe. In the USA, the initial concept was called *Building Product Models*, while in Europe it was presented as *Product Information Model* (EASTMAN *et al.*, 2011).

Despite the efforts of software manufacturers and organizations to promote BIM, the vast majority of projects are still developed using the traditional method, with 2D drawings and text documents. The contractor stands to gain the most from the adoption of BIM technology and, as the main stakeholder, should encourage the development of teams and the implementation of tools. (KYMMELL, 2008)

In countries in the northern hemisphere, the growth in the application of the BIM concept in architectural and engineering projects is evident, dealing in an integrated way with the elements of design, construction and management processes based on the formation of virtual models (FIESP. FEDERAÇÂO DAS INDÙSTRIAS DO ESTADO DE SÂO PAULO, 2008). International experience shows that this technology is being adopted in the AEC industry, improving productivity and increasing quality.

According to SOUZA *et al* (2009), encouraged by the infinite possibilities and facilities of BIM technology, some Brazilian offices followed the international movement, applying BIM systems in their companies in the early 2000s. This process has intensified in recent years, due to improvements in software and incentives to buy software. However, it was observed that they remained more on the shelves than they were actually used.

The lack of skilled labor, resistance to the introduction of new methodologies, the high cost of operation (investment in machinery and training), are some of the factors slowing down the complete implementation of the technology in the country's design offices.

As this is a new technology, the number of professionals who actually use BIM tools is still limited. In other words, anything new in a process brings risks and uncertainties, which tend to diminish with the passage of time and the spread of innovation. As new processes become more consolidated, uncertainty gives way to a more controlled and mature scenario (CHECCUCCI and AMORIM, 2011).

3.5 GOOD UNDERGRADUATE TEACHING

With the constant growth in the adoption of BIM by companies in the Architecture, Engineering and Construction (AEC) market, schools of Architecture and Civil Engineering have sought to implement subjects at undergraduate level to expose students to the challenges of parametric and collaborative project development (HOLLAND, 2010).

According to BARISON AND SANTOS (2011), from 2003 onwards, BIM teaching began to be included internationally in AEC courses, but this practice intensified between 2006 and 2009, related to strategies and approaches linked to the level of competence that the student must achieve in relation to the activity that will be carried out in professional practice.

3.5.1 INTERNATIONAL EXPERIENCE

According to CHECCUCCI (2014), there is no standardization in the adoption of BIM in teaching in other countries, and there is no consensus on the best way to do this. The author identified two basic ways of implementing the technology: by creating specific courses on the subject, or in existing courses. In this way, BIM is being inserted at different levels in undergraduate and postgraduate courses: through specializations, isolated or integrated subjects related to Architecture, Engineering and Construction Management.

In 2009, a survey was carried out in American universities, which found that: (1) in engineering courses, BIM is most widely used in the creation and design phase, through models, and is worked on, in the vast majority of schools, between the second and fourth years of the course and in a single specific subject; (2) the offer of BIM subjects occurred, for the most part, between the years 2006 and 2009, but teaching and learning advanced more quickly in architecture courses. (BECERICK-GERBER et al., 2011).

However, academic experiences dealing with BIM are new and based on pedagogies that have not yet been consolidated. According to SABONGI (2009), only 9% of North American construction schools teach BIM in their undergraduate courses, the main obstacles being: a lack of time or resources to redesign curricula, as well as a lack of university structure and prepared teachers, with specific materials related to teaching the technology.

WOO (2007), in his experience of teaching and learning BIM at the Department of Engineering Technology at Western Illinois University, using the Autodesk Revit Building 8.0 software, concluded that, in order to teach and learn BIM, one must have a vast knowledge of construction and that students with more experience in design and constructability, but less knowledge of computers, perform better when using the technology. In other words, it can be concluded that BIM should be started with simpler experiences, increasing the degree of difficulty and complexity of the tasks as the student's knowledge of buildings and construction increases.

Therefore, in order to develop a teaching strategy, universities are seeking, as the use of BIM technology spreads, some way of inserting it into the academic context, seeking to understand the role of BIM and how it has transformed the way of designing; that is, not limiting its use as just a "software package", but also as an exercise in collaboration, sustainable design, resource management, among others (ROMCY et al, 2013 *apud* KENSEK, 2012).

3.5.2 THE NATIONAL EXPERIENCE

The experiences related to teaching and learning BIM in Brazil show an evolving picture in the discussion of the subject in academic circles. Previously, the focus was on issues related to the three-dimensional and parametric model and its advantage over graphic representation in drawings. Today, there *is already* talk of collaborative and multidisciplinary work, and carrying out analysis with the model. Even so, the experiments were carried out on a one-off basis, in one or two subjects in the curriculum (CHECCUCCI, 2014).

The first approaches to BIM were made at the Mackenzie Presbyterian University in Sao Paulo, found in the work of VINCENT (2004), where students developed a multi-storey building using Autodesk Architectural Desktop software, through parametric modeling.

However, the first multidisciplinary experiment was carried out in the Civil Engineering and Architecture and Urbanism degree courses at the School of Civil Engineering of the State University of Campinas, where students learned how to insert the time factor (4D) into the project, as well as collaboration and interoperability. (RUSCHEL and GUIMARÂES FILHO, 2008)

RUSCHEL et al (2013), based on various didactic experiences researched, concluded that, on the national scene, BIM has been implemented gradually and with little effectiveness in Architecture, Engineering and Construction courses, and the main associated problems were the same as those highlighted by SABONGI (2009). It was noted that the international experience is much more mature than that found in Brazil, with BIM being addressed at various stages of the professional's training during the undergraduate course, which is also justified by a market demand in these countries, since the implementation of BIM by international companies has been happening more effectively and quickly.

Even though there is little national experience in teaching BIM, they have already made a positive contribution to its inclusion in undergraduate Civil Engineering teaching.

3.6 STRATEGIES FOR TEACHING BIM

For BARISON AND SANTOS (2011), "schools basically adopt two approaches: teaching BIM in one or two subjects or using BIM in several subjects in the curriculum. In the first approach, the BIM tool is usually taught in one subject at the beginning of the course and in another at the end of the course. In the second approach, the BIM model is used as a resource to help the student understand certain content."

After numerous studies at various American institutes, MOLAVI and SHAPOORIAN (2012) concluded that BIM should be included in the curriculum after students have a basic knowledge of construction management, and that the following order of content should be used (in a four-year course): (a) technical drawing, reading and architectural design; (b) AutoCAD for beginners (2D); (c) advanced AutoCAD (2D and 3D); (d) Revit Architecture; (e) advanced Revit; (f) BIM.

With regard to BIM teaching and learning strategies, BARISON AND SANTOS (2011), after a major review of international literature, classify the possibilities of working with BIM technology into three levels of competence that can be achieved by the training professional. This classification is related to the level of expertise that the student should have in professional practice using BIM, in relation to the complexity in which the concept is approached and, above all, whether its essential characteristics, such as interoperability, collaboration, communication and multidisciplinarity in project development, are actually addressed. In this way, the authors classify the courses into three different levels: introductory, intermediate and advanced.

1) Introductory level - covers the skills of the BIM MODELLER/FACILITATOR, teaching BIM tools, basic and fundamental concepts, basic study of modeling and how to communicate different types of information. Its purpose is to train students in modeling skills, extracting figures, altering models, creating components and the basic principles of communication and interoperability. It can be taught as part of a graphics course. Practical lessons on modeling a simple building and conceptualizing BIM are suggested. In this vein, the authors suggest that students modify a previously given model and then construct a small building.

2) Intermediate level - covers the skills of the BIM ANALYST, teaching other BIM tools, advanced 3D modeling techniques, different building systems and parametric object functionalities. The idea is to increase knowledge in modeling and advance in the creation of models and analysis of different building systems. To do this, students must work collaboratively and in teams, and must involve disciplines such as architecture, installations and structure. The work is done in groups, and it is up to each team to develop a complete multidisciplinary project, check for conflicts, analyze the model, raise costs and, above all, create collaboration and integration routines between the disciplines;

3) Advanced level - covers the skills of the BIM MANAGER, such as teaching BIM techniques and processes related to interoperability, management concepts and tools, as well as implementation and team dynamics. To do this, the student must have knowledge of the main types of BIM tools, building materials, construction technologies and the *modus operandi* of the construction industry. The focus at this stage is on helping the student to understand how BIM can help construction management techniques. At this level, the focus should be on forming multidisciplinary teams that work in a synchronized manner and, above all, on using management tools integrated with the construction information model, optimizing construction management with 4D (time) and 5D (cost) simulations. In order to achieve good results at the advanced level, it is suggested, despite the difficulties, to integrate with projects being developed in the labor market, thus following the development of the project from conception, conflicts and construction management. In this way, the authors believe it is possible to integrate students directly with the job market.

These same authors believe that a partnership with companies in the AEC industry is essential for combining theoretical and practical knowledge.

CHECCUCCI (2014) believes that integrated practice is the most appropriate way to teach BIM in schools, as this approach allows students to study the technology within the specific contexts of each subject, allowing them to discuss the various interfaces of BIM with teachers who are specialists in different areas of knowledge. For her, work using a single BIM model in one or more disciplines, as long as it focuses on different stages of a building's life cycle, getting a group of teachers to work on different aspects of modeling are some possible approach strategies.

According to the same author, this integrated strategy has several advantages such as: (a) less change in the curriculum structure, since the time required will be distributed among the different subjects in the curriculum; (b) the possibility for students to work on the different specialties with different teachers, and to have access to different ways of using Building Information Modeling.

According to CES/CNE Resolution 11 (MEC, 2002), a reference for the preparation of curricula for undergraduate engineering courses, including Civil Engineering, the curriculum is divided into three levels: (i) the basic content core, which accounts for around 30% of the minimum workload; (ii) the professional content core, which accounts for around 15% of the minimum workload and (iii) the specific content core, made up of extension activities and in-depth study of the content in the professional content core, which accounts for the rest of the total workload. As it is a set of subjects and/or activities proposed exclusively by the HEI, it covers the scientific, technological and instrumental knowledge needed to define the types of engineering and must guarantee the development of the competences and skills established in these guidelines. Although BIM-related content is not provided for in this resolution, some Civil Engineering courses have approached the subject through research at undergraduate and postgraduate level, as well as experiences in the use of BIM in project development disciplines.

4 METHODOLOGY

This project was developed in three stages. In the first stage, a bibliographical survey was carried out, covering the following subjects: (a) The conceptualization of BIM, as well as its characteristics and advantages; (b) The evolution of graphic representation technologies in a historical context; (c) The approach to the BIM platform in teaching and the job market, with a greater focus on education, both in the national and international scenarios; (d) Research in the area of civil engineering teaching integrated with BIM; (e) Pedagogical projects of undergraduate Civil Engineering courses at Brazilian universities.

In the second stage, data was collected by means of a questionnaire on the academic community's knowledge of BIM.

Finally, a qualitative analysis was made of the answers obtained by applying the questionnaires, considering the feasibility of inserting BIM concepts into Design and Project disciplines in an integrated and multidisciplinary way.

As a final part of this research, a questionnaire was drawn up in order to identify how in-depth the knowledge of the teaching staff of Brazilian universities is regarding Building Information Modeling.

This questionnaire was sent via the web to approximately 400 professors of Civil Engineering courses at 42 public universities and 14 private universities, using the e-mail addresses provided on the departments' websites. The computer platform used to send the questionnaire was Google Docs®, as it offers the option of sending it via e-mail or accessing it via a link. It is a free, easy and fast to use platform, whose data is stored in a spreadsheet, automatically generating a summary containing the absolute values, percentages and quantitative graphs of the answers. This questionnaire was available to answer for three months.

The questionnaire used to gather the data for this survey initially included a series of questions designed to get to know the respondent's profile. This stage of the questionnaire included questions about the university where the respondent works, the area in which they work, the length of their training and the length of their teaching experience. Afterwards, the respondent was asked if they used BIM technology in their teaching activity. If the answer was yes, a series of questions began about how long they had been using the BIM tool, their level of experience and stage of application in the use of modeling in their teaching activities and the changes that were most noticeable after the insertion of BIM in their teaching activity; if not, they went straight to more general questions on the subject, taking into account their academic experience. Questions about possible obstacles to the implementation of technology in education formed part of this stage of the questionnaire. Finally, we asked at what point in the Civil Engineering degree course the lecturer thought the possible inclusion of BIM was most coherent and what aspects of the training of AEC professionals the lecturer thought were relevant in the context of BIM.

The questionnaire sent out can be found in Annex 1 at the end of this report.

5 RESULTS AND DISCUSSION

The results of the questionnaire survey are based on 48 complete answers, which represent 12% of the sample of 400 teachers who were sent the questionnaires. Of the respondents, the majority were from public universities (89.6%). This is due to the fact that the vast majority of the questionnaires were sent to teachers at public universities, where there is a greater concentration of research and studies in a wide variety of areas.

As for the profile of the respondents, 81% answered that they had more than 15 years of undergraduate teaching experience, with no responses indicating less than 5 years of teaching experience. In terms of total teaching experience, 62% have more than 15 years' teaching experience, compared to 17% with between 1 and 5 years' experience in the academic field. By relating these two issues, we can observe a common profile to be found in our country's universities: the majority of lecturers have a relatively long period of training, as well as teaching experience. This is due to the fact that, in order to become a lecturer in universities, people have to go through the postgraduate levels - masters and doctorate, which take around 6 years to complete. It's rare to find teachers who are very young or who have only been trained for a short time, even in private universities. Figure 2 and Figure 3 show the distribution of the results relating to the profile of the respondent.

Figure 2 - Time since Graduation

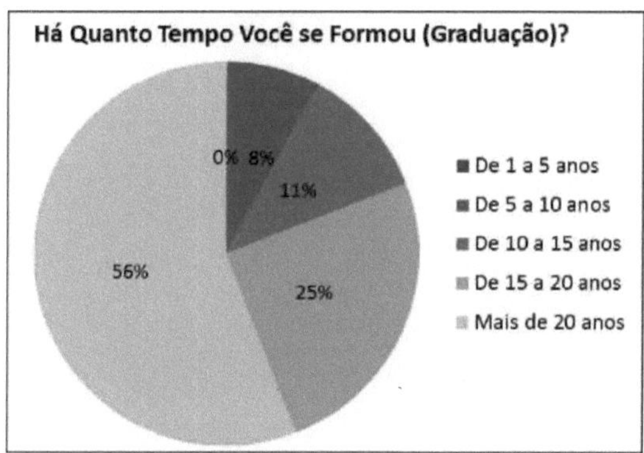

Figure 3 - Length of Teaching Experience

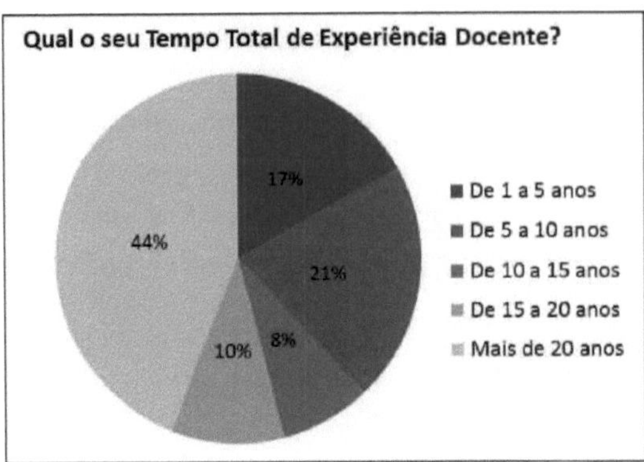

With regard to the area in which they work, Figure 4 shows that the majority are in the field of Civil Construction (35%), followed by Hydraulics, Sanitation and the Environment (19%); Geotechnics (6%); Transportation (10%); Structures (25%) and Architecture and Urban Planning (4%). At first glance, the low percentage of professors working in the field of Architecture and Urbanism may seem a little strange, as it is the field that is currently most likely to implement BIM compared to the other areas. However, the explanation for this is due to the fact that the number of professors working in this field is low in undergraduate Civil Engineering courses, justified by the reduced number of hours of subjects linked to this field, in contrast to the Civil Construction area, whose subjects were at the heart of the course.

Figure 4 - Area of Operation

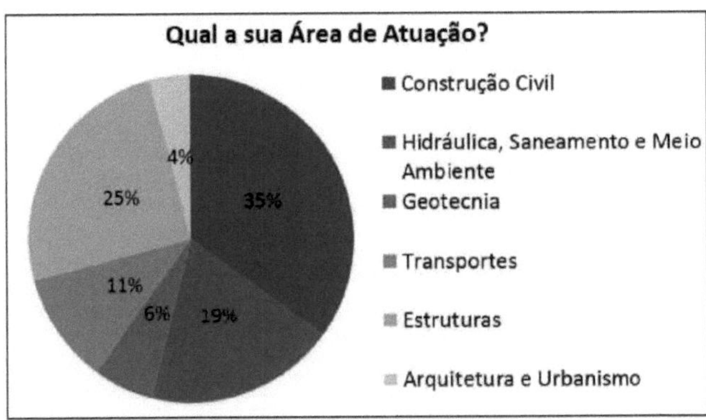

Regarding the use of BIM technology in teaching, 73% said they did not use it. Of the 27% who said yes, 83% had been using it for less than 5 years. In other words, we can see that BIM technology is still not very widespread in academia and, as a result, its use is still very restricted. This is also due to the fact that it is a recent technology, about which knowledge is still in its infancy. Figure 5 and Figure 6 show this picture.

Figure 5 - Use of BIM in teaching activities

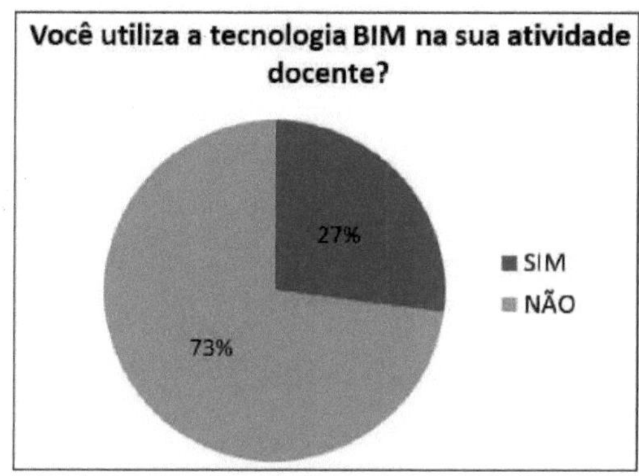

Figura 6 - Time spent using the BIM tool

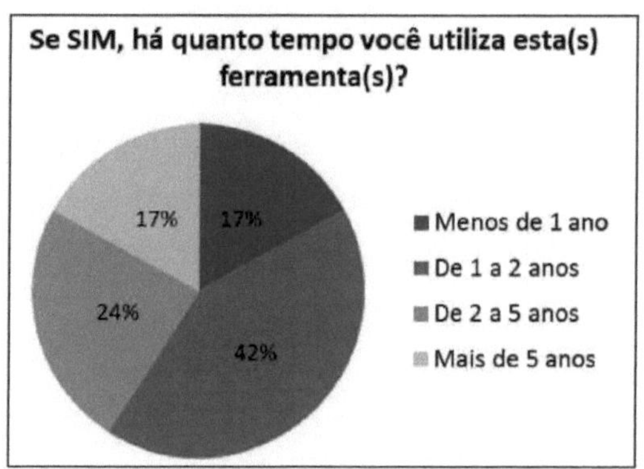

In order to relate the experience of using the BIM tool in teaching activities to the various academic levels, we asked about activities in the undergraduate and postgraduate areas, subdividing them into teaching and research activities, as well as extension. Figure 7 shows the results. In terms of undergraduate teaching, the BIM tool is still little covered in the majority of activities, with only 20% of responses indicating activities with practical and theoretical content on the subject, and 24% not covering BIM in their area of work. In research activities at undergraduate level, 14% have the BIM tool as the main theme of the activity. Looking at postgraduate courses, the same scenario can be observed as in undergraduate courses, as the majority of respondents indicate that they do not address BIM technology in their academic activity. The use of BIM is even more restricted in extension activities, where only 5% use the tool as the main theme of the activity.

Figura 7 - Experience of using BIM in teaching

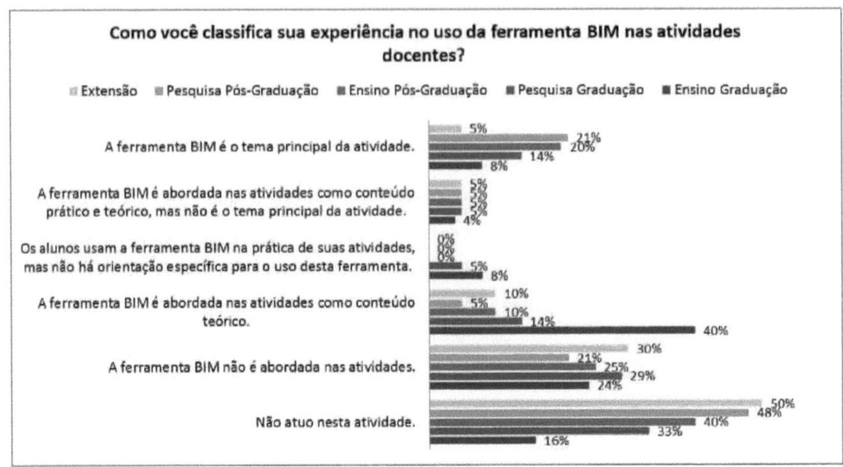

For those who approach BIM in their academic activities, according to the stages defined by RUSCHEL *et al* (2011), TOBIN (2008) and SUCCAR (2009), the results obtained suggest that the main focus of the BIM approach is in Stage 1 (58%), characterized by parametric modeling, followed by activities within Stage 2 (26%), focusing on interoperability and asynchronous collaboration, and finally Stage 3 (16%), with synchronous integrated practice. We can see this result in Figure 8 below.

Figura 8 - BIM application stage

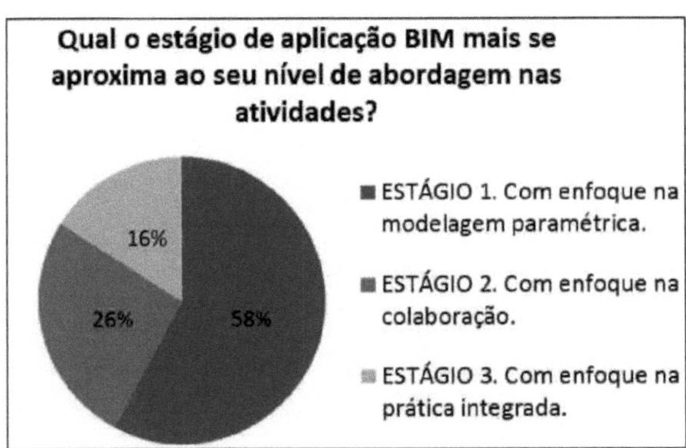

The questionnaire listed some of the possible beneficial changes that the insertion of BIM would bring with its use, so that the teachers could analyze them and rate them on a scale of 1 to 5, where 1 represents no change and 5 major changes, according to their experience. We can see the result in **Error! Reference source not found**. 63% were able to see advances in anticipating design problems and making them easier to visualize; 57% said there had been major improvements in the richness of design details; 37% saw low to medium impact changes in terms of improving compatibility between designs and 47% said there had been major improvements in reducing graphical representation errors. In this way, we can see that, in all the points analyzed, improvements in project development were indicated with the use of the tool. In addition, a space was opened up for teachers to add other changes they considered important, and among the changes listed, the most cited were improved teamwork and an integrated view of the project's life cycle using BIM technology.

Figura 9 - Possible Changes with the Use of BIM

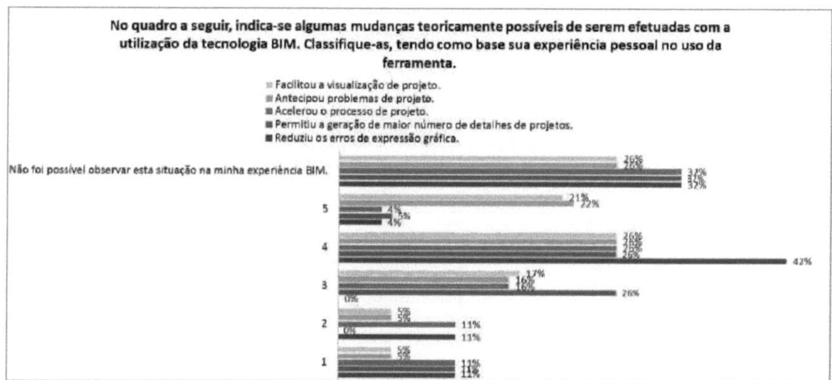

Specifically regarding the implementation of BIM technology in the academic area, some aspects are considered influential, and these were classified by the respondents on a scale of 1 to 5, where 1 represents not very influential and 5 very influential. The most influential aspect for the implementation of BIM technology was the availability of BIM teaching infrastructure in universities, such as laboratories, hardware and software, as 94% of the responses indicate that this is a highly important issue in the implementation of BIM in the academic area. The next most influential aspects were greater integration between course subjects (81%), compatibility with CAD tools that have already been consolidated (79%), access to teaching materials (73%) and resistance to adopting BIM, as it is a technology that has not yet been consolidated in the academic world (69%). The respondents were also able to add other influential aspects to the implementation of BIM in universities, including the need to engage the teaching staff in changing methodology and specific standards on the subject. Figure 10 shows this picture.

Figura 10 - Influential Aspects in the Academic Area

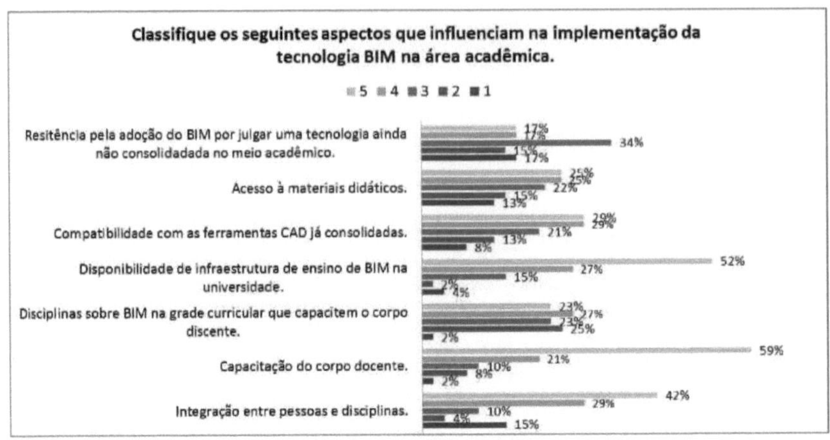

Finally, in relation to the best time in the Civil Engineering course curriculum for BIM content to be adopted, based on Resolution CES/CNE 11 (MEC, 2002), the following responses were obtained, as shown in Figure 11 below: 48% believe that the implementation of BIM should be at the level of specific content; 41% at the level of vocational content and 10% at the level of basic content. In addition to this question, some lecturers stressed that BIM content should be included at more than one point in the Civil Engineering course, while others believe that it should be covered in an informative way, playing a supporting role in the Civil Engineering teaching process.

Figure 11 - Best Time to Insert BIM

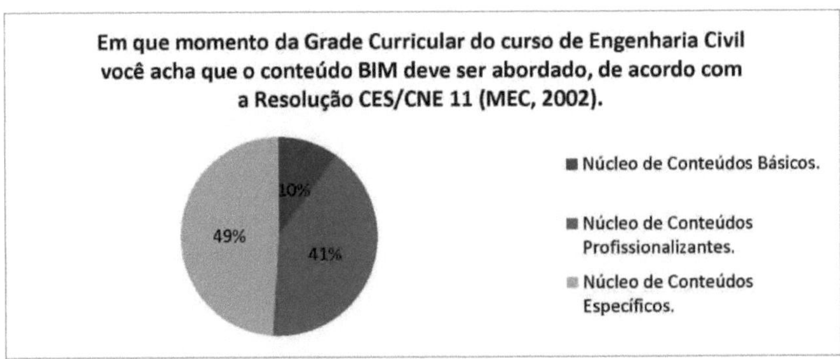

6 CONCLUSIONS

According to the data obtained by applying the questionnaires, it is possible to observe that knowledge of BIM technology is still restricted, even in the academic field, where there is a greater chance of approaching new technologies and new horizons of knowledge. This is due to the fact that the BIM paradigm is recent and has not yet been consolidated, and is being inserted into both education and the job market at a slow and conservative pace in Civil Engineering (RUSCHEL *et al*, 2013). This reality is directly linked to the academic community's resistance to change and their lack of knowledge of how to use BIM technology.

In order to implement BIM teaching in the academic environment, the infrastructure of universities is a crucial point, since it was the aspect considered most relevant by the teachers who answered the questionnaire. For this reason, one of the major obstacles to the introduction of BIM in Civil Engineering courses is the lack of structure in terms of physical training, such as laboratories, hardware and software, and the professional training of teachers and technicians.

Given this scenario, when thinking about when BIM concepts should be included in Civil Engineering courses, first of all, according to BARISON and SANTOS (2011), we need to know what levels of BIM knowledge the student should have, and how much time will be allocated in the curriculum to teaching this subject. Once this aspect has been defined, one strategy pointed out in the results is to include BIM in the three levels of the course curriculum, with each approach being compatible with the student's level of education and training needs.

In this way, BIM, at an introductory level, should be included in the basic cycle of the Civil Engineering course curriculum, associated with Graphic Expression subjects such as Computer Aided Design and Technical Drawing, including an initial approach to the concept of BIM and parametric modeling, equivalent to Stage 1. BIM can also be introduced in the vocational content core, since this stage of the course requires the student to be more knowledgeable about construction materials and methods, investing in knowledge about modeling, teamwork, conflict analysis and integration between disciplines, in an approach equivalent to Stage 2. Finally, at the end of the degree course, it would be important to take an advanced approach to BIM, equivalent to Stage 3, exploring the principles of interoperability, management, project conflicts and simulations in 4D (which considers, in addition to project dimensions, the variable "time" in modeling) and 5D (which considers the variable "cost" in the project, in addition to dimensions and time). As this is a gradual and evolutionary process, each level should be introduced according to the availability of resources at universities, as well as the faculty's knowledge of the subject.

Finally, this work, together with the others analyzed in the literature review, supports the teaching of

BIM as an essential method for training professionals in Civil Engineering. It is believed that knowledge of this technology will soon be a decisive factor in hiring by AEC companies, so it is important that universities are aligned with the demand for knowledge required by the job market and, in this way, repair their teaching staff, students and their physical and curricular structure to achieve these objectives.

7 BIBLIOGRAPHICAL REFERENCES

ANDRADE, M. L. V. X. de. **Performative Design in Recent Architectural Practice: Conceptual Framework**. 2012. 472p. Thesis (Doctorate in Architecture and Urbanism) - Faculty of Civil Engineering, Architecture and Urbanism, State University of Campinas, Campinas, Sao Paulo, Brazil.

BARISON, M. B. and SANTOS, E.T. **Current Trends in BIM Teaching.** V TIC - Meeting of Information and Communication Technology in Construction. Salvador, Brazil. 2011.

BECERIK-GERBER, B.; GERBER, D. J.; KU, K. The pace of technological innovation in architecture, engineering, and construction education: integrating recent trends into the curricula. Journal of Information Technology in Construction (ITcon), v. 16, p. 411432, 2011. Available at: <http://www.itcon.org/2011/24>. Accessed on: Feb. 14, 2013.

CEROVSEK, T. A Review and Outlook for a 'Building Information Model' (BIM): A Multi-Standpoint Framework for Technological Development. Advanced Engineering Informatics. 25. Ed. p.224-244. 2011.

CERVO, A. L. e BERVIAN, P. A. **Metodologia Cientifica**. 5. ed. Sao Paulo: Prentice Hall. 2002. 242 p.

CHECCUCCI, E. S. **Teaching-Learning of BIM in Undergraduate Civil Engineering Courses and the Role of Graphic Expression in this Context.** Thesis (Multi-Institutional and Multidisciplinary Doctorate in Knowledge Dissemination). Salvador, Brazil. 2014

CHECCUCCI, E. S.; AMORIM, A. L. **Construction Information Modeling as Technological Innovation.** V TIC - Meeting of Information and Communication Technology in Construction. Salvador, 2011.

COELHO, S. S. and NOVAES, C. C. **Building Information Modeling (BIM) and Collaborative Environments for Project Management in Civil Construction.** 2008. Available at: <www2.pelotas.ifsul.edu.br/gpacc/BIM/referencias/

COELHO_2008.pdf> . Accessed on: May 2013.

CONFEA. Resolution No. 1010, of August 22, 2005. Provides for the Regulation of the Attribution of Professional Titles, Activities, Competencies and Characterization of the Scope of Practice of Professionals Inserted in the CONFEA/CREA System, for the purpose of supervising Professional Practice. Available at: <www.confea.org.br/media/res1010.pdf> Accessed in: May 2013.

EASTMAN, C. et al. BIM Handbook: A Guide to Building Information Modeling for Owners, Managers, Designers, Engineers and Contractors. 2. Ed. New Jersey: John Wiley & Sons Inc. 2011.

FU, C.; et al. **IFC Model Viewer to Support nD Model Application.** Automation in Construction. v.15. p.178-185. 2006.

HABITARE. **Innovation, Quality & Productivity Management and Knowledge Dissemination in Housing Construction**. Vol. 2. 2003 a. Available at: <http://www.habitare.org.br/publicacao coletanea2.aspx>. Accessed on: October 20, 2008.

HOLLAND, R.; MESSNER, J.; PARFITT, K.; POERSCHKE, U.; PIHLAK, M.; SOLNOSKY, R. **Integrated Design Courses Using BIM as the Technology Platform.** Available at: <http://php.scripts.psu.edu/users/r/l/rls5008/CV/Linked%20Papers/Holland%20et%20al %202010%20Integrated%20Design%20Courses%20Using%20BIM%20as%20the%20 Technology%20Platform.pdf>. Accessed on: May 10, 2014.

. **Standardization and Certification in Housing Construction**. Vol. 3. 2003b. Available at: <http://www.habitare.org.br/publicacao coletanea3.aspx>. Accessed on: 20. Oct.2008.

JERNIGAN, F. Big BIM, Little BIM: The Practical Approach to Building Information Modeling integrated practice done the right way! Salisbury: 4 Site Press, 2007. 323 p.

KENSEK, K. M. Advancing BIM in Academia: Explorations in Curricular Integration. Computational Design Methods and Technologies: Applications in CAD, CAM and CAE Education. IGI Global, 2012.

KYMMELL, W. Building Information Modeling: Planning and Managing Construction Projects with 4D CAD and Simulations. New York: McGraw Hill. 2008.

LEE, Seung-Il et. al. Efficiency Analysis of Set-based Design with structural building information modeling (S-BIM) on high-rise building structures. Automation in Construction. 23. Ed. p. 20-32. 2012.

MEC. Ministry of Education. CNE/CES Resolution 11, of March 11, 2002. National Guidelines for Engineering Degree Courses. Available at: < http://portal.mec.gov.br/cne/arquivos/pdf/CES112002.pdf>. Accessed on: May 2014.

MOLAVI, J.; SHAPOORIAN, B. **Implementing an interactive program of BIM applications for graduating students.** International Conference on Sustainable Design, Engineering and Construction. 2012. Fort Worth: ASCE, 2013. p. 1009-1016. Available at: <http://cedb.asce.org/cgi/WWWdisplay.cgi?298975>. Accessed on: July 26, 2014.

MONTEIRO, A. **Design for the production of vertical masonry walls using a CAD-BIM tool**. 2011. Dissertation (Master's Degree in Civil Engineering) - Department of Civil Construction Engineering, Polytechnic School of the University of Sao Paulo. Sao Paulo, Brazil.

OLIVEIRA, S. L de. Tratado de Metodologia Cientifica: Projetos de Pesquisas, TGI, TCC, Monografias, Dissertaçoes e Teses. 2. Ed. Sao Paulo: Pioneira Thomson Learning. 2001. 320p.

PENTTILLÂ, H. Describing the Changes in Architectural Information Technology to Understand Design Complexity and Free-form Architectural Expression. Itcon. V.11.2006

PIMENTEL, J. P. Survey and Characterization of BIM Tools Applied in the Rationalization of Building Systems Projects. PIBIC/CNPq/UFSCar Scientific Initiation Final Report. Sao Carlos. 2012.

PORWAL, A.; HEWAGE, K. N. Building Information Modeling (BIM) partnering framework for public construction projects. Automation in Construction. 31. Ed. p. 204-214. 2013

ROBSON, C. Real World Research: A Resource for Social Scientists and PractitionerResearchers. 2. Ed. Willey. 2002. 624p.

ROMCY, N. M. S.; CARDOSO, D. R.; MIRANDA, N. M. **BIM e Ensino: Experiência Acadêmica Realizada na Universidade Federal do Cearâ**. VI TIC - Meeting of Information and Communication Technology in Construction. Campinas, Brazil. 2013.

RUSCHEL, R. C.; ANDRADE, M. L. V. X.; SALES, A. A.; MORAIS, M**. Teaching BIM: Examples of Implementation in Architecture Engineering Courses**. V TIC - Meeting of Information and Communication Technology in Construction. Salvador, 2011.

RUSCHEL, R. C.; ANDRADE, M. L. V. X. e MORAIS, M**. O Ensino de BIM: onde estamos**. Built Environment. V.13. Porto Alegre, 2013.

RUSCHEL, R. C.; GUIMARÂES FILHO, A. B. **Getting started with CAD 4D**. Brazilian Workshop on Project Process Management in Building Construction. Sao Paulo: USP, 2008. SABINO, W. CAD-BIM-GIS: Integrating Solutions. 2008. Available at: http://pasta.ebah.com.br/download/cad-bim-gis-integracao-de-solucoes-wolmar-sabino- ppt-7866 Accessed on: October 20, 2008.

SABONGI, F. J. The Integration of BIM in the Undergraduate Curriculum: an analysis of undergraduate courses. In: 45th Annual Conference of Associated Schools of Constructio, 2009. Available at: <http://ascpro0.ascweb.org/archives/cd/2009/paper/CEUE90002009.pdf> Accessed on: May 28, 2014.

SACKS, R. et. al. Requirements for building information modeling based lean production management systems for construction .Automation in Construction. 19. Ed. p. 641-655. 2010.

SANTOS, E. T.; BARISON, M. B. **BIM and universities**. Construçao e Mercado magazine. Available at: <http://revista.construcaomercado.com.br/negocios- incorporacao-construcao/115/o-desafio-para-as-universidades-formacao-de-recursos- humanos-208417-1.asp>. Accessed on: May

2013.

SMITH, D. K. and TARDIF, M. Building Information Modeling: A Strategic Implementation Guide for Architects, Engineers, Constructors, and Real Estate Asset Managers. New Jersey: John Wiley & Sons Inc. 2009.

SOUZA, L. L. A.; AMORIM, S. R.; LYRIO, A. M. **Impacts of using BIM in Architecture Offices: Opportunities in the Real Estate Market.** Project Management & Technology. v.2. Sao Paulo, Brazil. 2009.

SUCCAR, B. Building Information Modeling Framework: A research and delivery foundation for industry stakeholders. Automation in Construction. 18. Ed. p. 357-375. 2009.

TSE, T. C.; WONG, K. D. and WONG, K. W. **The Utilization of Building Information Models in nD Modeling: A Study of Data Interfacing and Adoption Barriers.** Journal of Information Technology in Construction. v. 10. p. 65-110. 2005.

TOBIN, J. **Building Information Modeling for tertiary construction education in Hong Kong**. Journal of Information Technology in Construction. Sept. 2011. Available at: <http://www.itcon.org/cgi-bin/works/Show?2011_27> Accessed on: May 10, 2014.

VINCENT, C. C. **Architectural design and computer graphics: processes, methods and teaching**. Congress of the Ibero-American Society of Digital Graphics, 8., 2004, Sao Leopoldo. p. 89-90. Available at: <http://followscience.com/content/525081/projeto- arquitetonico-e-computacao-grafica-processos-metodos-e-ensino>. Accessed on: September 13, 2013.

WOO, J. H. **BIM (Building Information Modeling) and Pedagogical Challenges**. Annual Conference by associated Schools of Construction, 43., 2007, Arizona. Proceedings ... Arizona: SULBARAN, T.; CUMMINGS, G., 2007. Available at: <http://ascpro0.ascweb.org/archives/cd/2007/paper/CEUE169002007.pdf>. Accessed on: Feb. 14, 2013.

ZHANG, S. et. al. Building Information Modeling (BIM) and Safety: Automatic Safety Checking of Construction Models and Schedules. Automation in Construction. 29. ed. p. 183-195. 2013.

8 TECHNICAL AND SCIENTIFIC PRODUCTION

The article based on the results of this work was approved for the XLII Brazilian Congress of Engineering Education (Engineering: Multiple Knowledge and Actions), which will be held from September 16 to 19, 2014 in Juiz de Fora, Minas Gerais. It was selected to be presented in the Technical Section. This article is in Annex 2 of this document.

9 SELF-EVALUATION

Undertaking the Scientific Initiation project was an incredible experience, both personally and professionally. Since the beginning of my degree, I had been waiting to find the right moment, the right supervisor and the perfect topic. And it couldn't have been any different, or even better.

Studying, understanding and learning BIM, in its many interfaces and applications, was essential, since it is a methodology that is becoming more and more widespread on the national scene, and especially on the international scene, as well as being the future of engineering. In a short time, knowing BIM will be a fundamental prerequisite for access to the job market, especially in large companies in the Architecture, Engineering and Construction (AEC) sector, because it reduces costs, time and optimizes performance, compatibility between projects and graphic representation.

Through this research, I developed the skills and attitudes of a researcher, such as working in partnership, discipline, commitment and dedication, as well as learning how to produce technical-scientific texts and how to structure them. On the other hand, continuing this work has required me to better organize my time personally, combining research and my degree, in order to fulfill my duties and achieve the best possible results.

Finally, I would like to say a huge thank you to my great advisor and friend, Rochele Amorim Ribeiro. She was the one who taught me how to crawl along the path of research, and who helped me so much that this work was successful and had satisfactory results.

And God, thank You for Your hand outstretched over me, both in times of trouble and in times of joy. I owe everything I have to You.

Pedro Augusto Izidoro Pereira

10 ADVISOR EVALUATION

Student Pedro did an excellent job of research, showing interest, assiduousness and responsibility in the development of the work. He also showed proactive and creative behavior in solving problems that arose during the course of the work, especially during the survey and data collection stage using questionnaires. The scholar has kept to the research timetable as laid out in the plan, managing to overcome the difficulties and challenges inherent in the research topic.

In this way, I rate Pedro's performance in this Scientific Initiation work as GREAT.

....

Rochele Amorim Ribeiro

11 STUDENT DESTINATION

The scholar will continue his degree in Civil Engineering at the University - UFSCar, entering the tenth - and last - period of the course, so it will not be possible for the scholar to continue this work in other scientific initiations. However, the topic will be developed in future research projects submitted by the supervising professor.

12 ANNEXES

12.1 ANNEX1: Questionnaire Applied to Teachers

QUESTION 1 Which university are you currently working at?

QUESTION 2: How long ago did you graduate?

a) From 1 to 5 years

b) From 5 to 10 years

c) From 10 to 15 years old

d) From 15 to 20 years old

e) More than 20 years

QUESTION 3. What is your total teaching experience? Including experience prior to this University.

a) From 1 to 5 years

b) From 5 to 10 years

c) From 10 to 15 years old

d) From 15 to 20 years old

e) More than 20 years

QUESTION 4. What area do you work in?

a) Construction

b) Hydraulics, Sanitation and the Environment

c) Geotechnics

d) Transport

e) Structures

f) Architecture and Urban Planning

QUESTION 5: Do you use BIM technology in your teaching? If NO, please go to question 10.

a) YES

b) NO

QUESTION 6. If YES, how long have you been using these tool(s)?

a) Less than 1 year

b) From 1 to 2 years

c) From 2 to 5 years

d) More than 5 years

QUESTION 7. How would you rate your experience of using the BIM tool in the following teaching activities? (Each teaching activity should be rated with each of the options).

Teaching Activities:

1) Undergraduate Education

2) Undergraduate Research

3) Postgraduate Education

4) Postgraduate Research

5) Extension

Ratings:

a) I don't work in this field.

b) The BIM tool is not covered in the activities.

c) The BIM tool is covered in the activities as theoretical content.

d) The students use the BIM tool in their work, but there is no specific guidance on how to use it.

e) The BIM tool is covered in the activities as theoretical and practical content, but it is not the main theme of the activity.

f) The BIM tool is the main theme of the activity.

QUESTION 8. Which stage of BIM application described in the alternatives below comes closest to your level of approach in your activities?

a) STAGE 1. With a focus on parametric modeling, technology is used only as a specific tool, because the work process is still individualized, without the involvement and collaboration of other disciplines. It is generally restricted to a specific phase of the process (design, construction or operation).

b) STAGE 2: With a focus on collaboration, the technology is expanded to professionals from other disciplines (multidisciplinary between one or two phases of the design process, involving up to two different agents, such as architecture and structure, or cost management) and interoperability emerges as an essential term for BIM to become a working process. This stage is characterized by the manipulation of a single three-dimensional model between all the teams involved in the work. The process is also interactive and still asynchronous, but with improved interoperability between the agents involved.

c) STAGE 3. With a focus on integrated practice (shared and collaborative creation of the building model, throughout the development process, involving the design, construction and operation phases, and the multiple disciplines of the AEC area), it expands the possibilities of interoperability through open protocols and virtual work environments, so that all the agents involved in the development can contribute collectively, within the specificities of their disciplines. This stage is characterized by network integration, whose process is synchronous, involving complex analyses already in the early stages of conception, such as the analysis of design conflicts.

QUESTION 9. The table below shows some changes that could theoretically be made using BIM technology. Please rate them, based on your personal experience of using this tool. The rating will be given on a scale of 1 to 5, where 1 means that nothing has changed and 5 means that a lot has changed. Give each item a score according to your experience with the technology.

a) Improved compatibility between projects.

b) Reduced graphical representation errors.

c) It allowed the generation of a greater number of project details.

d) It accelerated the design process.

e) Anticipated design problems.

f) It made it easier to visualize the project.

Write below if you have noticed any other changes when using BIM technology.

QUESTION 10. Rate the following aspects that influence the implementation of BIM technology in the academic field. The rating will be given as follows: a scale of 1 to 5, where 1 represents an aspect of little influence and 5 represents an aspect of great influence.

a) Integration between the parts that make up the work team (people and disciplines). b) Training of the teaching staff.

c) Subjects on BIM in the curriculum that train the student body.

d) Availability of BIM teaching infrastructure at the university (equipment, software, laboratories).

e) Compatibility with established CAD tools.

f) Access to teaching materials: books, tutorials, video lessons.

g) Resistance to adopting BIM for teaching/research, as it is a technology that has not yet been consolidated in the academic world.

If there are other aspects that you think are influential in the implementation of BIM, what are they?

QUESTION 11. At what point in the Civil Engineering course curriculum do you think BIM content should be covered (according to RESOLUTION CES/CNE 11 (MEC, 2002))? Check as many boxes as you think is coherent.

a) Core Subjects. Aimed at acquiring general knowledge about engineering and its basic sciences (Physics, Mathematics and Chemistry), this corresponds to around 30% of the minimum course load.

b) Professionalizing Content. Corresponds to 15% of the minimum course load and has a coherent subset of topics, which may or may not be specific to Civil Engineering (e.g. Civil Construction, Environmental Management...).

c) Specific Content Core. This consists of extensions and deepening of the contents of the core of professionalizing contents, as well as other contents intended to characterize modalities. The specific, technological and instrumental knowledge required to define engineering disciplines must guarantee the development of the competences and skills established in these guidelines. The workload is proposed by the HEI itself and can cover up to 55% of the minimum workload.

If you have anything to add to the above question, this is the place.

QUESTION 12: What aspects of the training of professionals in the field of architecture, engineering and construction do you consider relevant in the context of BIM?

12.2 ANNEX 2: Article Approved for COBENGE

THE INCLUSION OF BIM IN THE UNDERGRADUATE COURSE IN CIVIL ENGINEERING

Pedro A. I. Pereira - pedroaipereira@gmail.com

Rochele A. Ribeiro - rochele@ufscar.br

Federal University of Sao Carlos - Department of Civil Engineering

Washington Luis Highway, km 235 -SP310

ZIP Code 13565-905 - Sao Carlos - Sao Paulo

Abstract: This article investigates the current conditions of civil engineering students' university education with regard to training in the use of the BIM (Building Information Modeling) platform on the national scene. Through bibliographical research and the evaluation of questionnaires answered by professors from Brazilian universities, the main objective of this research was to identify pedagogical strategies for approaching the BIM platform in undergraduate Civil Engineering teaching, taking into account CES/CNE Resolution 11 (MEC, 2002). The results point to the insertion of BIM in the three core areas of the course curriculum as a strategy, with each approach being compatible with the student's level of education and training needs. Finally, this work, together with the others analyzed in the literature review, supports the teaching of BIM as an essential subject for the training of Civil Engineering professionals.

Keywords: BIM; Undergraduate Teaching; Civil Engineering.

1. INTRODUCTION

The successful development of building and construction projects depends above all on the ability to reconcile different pieces of information about the elements involved in the process. The development of the project involves various constraints, such as information on material specifications, labor costs, execution time, compatibility of infrastructure installations, among others, which need to be handled simultaneously in the planning and execution process. In this way, the civil engineering professional must integrate all this information in a systematic and accurate way in order to make the best planning decisions and comply with the proposed construction schedule (HABITARE, 2003a; 2003b).

Civil Engineering therefore uses computer tools to optimize the process of preparing building projects, with *Computer-Aided Design* (CAD) being the main exponent. However, due to the increasing complexity of construction projects, it is essential to aggregate information in alphanumeric data (costs, quantity of material, quality of material, for example) in the spatial entity, enabling integrated management of this information. One solution to this problem is to make use of Building *Information Modeling*, known as BIM.

Unlike the 3D model in CAD language, which only represents the three-dimensional features of the building elements, the BIM model allows the association of information about the components that make up these elements. In other words, in the BIM model, the building elements are linked to graphic, three-dimensional, quantitative and parametric attributes, allowing for the generation of descriptive documents about the work, such as two-dimensional representations (e.g. floor plans, sections, façades, details), material performance analysis (e.g. acoustics, thermal comfort), budget spreadsheets and physical and financial schedules. In addition, using the BIM platform, it is possible to develop a collaborative project, which allows the intervention of several teams responsible for the project, considering the updates of the project in a consistent, non-redundant and synchronous way (SMITH and TARDIV, 2009; CEROVSEK, 2011; EASTMAN et. al., 2011).

The training of Civil Engineering professionals in the use of the BIM platform is a growing concern in undergraduate courses and in university research. It is more evident on the international scene, but it is growing on the national scene. Consequently, there is a need for a pedagogical strategy that can better integrate knowledge of the application of the BIM tool with the teaching needs of Civil Engineering Design in all areas: architectural, structural, infrastructure, among others.

Therefore, this article investigates the current conditions of university education for Civil Engineering students with regard to training in the use of the BIM platform on the national scene. Through bibliographical research and the evaluation of questionnaires answered by professors at Brazilian universities, the main objective of this research was to identify pedagogical strategies for approaching the BIM platform in undergraduate Civil Engineering teaching. Secondary objectives were: (i) to survey scientific research that addresses the issue of BIM in undergraduate teaching; (ii) to survey the profile of the teaching staff of undergraduate courses that work with BIM, with an emphasis on the national scenario; (iii) to survey the possibilities for integrated BIM in the disciplines of undergraduate Civil Engineering courses, taking into account CES/CNE Resolution 11 (MEC, 2002). It is hoped that the results of this research will help to define pedagogical strategies to enable the inclusion of the BIM theme in undergraduate Civil Engineering teaching, by means of an integrated and multidisciplinary approach to BIM in the course curriculum.

2. CLASSIFICATION OF THE STAGES IN THE USE OF BIM

The inclusion of the BIM platform in the development and documentation of projects in engineering and architecture is a growing process in the global context, although there are different stages of adoption of the platform. According to PORWAL and HEWAGE (2013), four levels of maturity have been identified in the use of computational tools for project development: (1) Level zero, characterized by the use of CAD2D as tools for two-dimensional entities, such as lines, arcs, texts, among others; (2) Level one, characterized by the use of 2D and 3D CAD simultaneously, with the elaboration of three-dimensional models; (3) Level two, such as the use of BIM, which allows the parameterization of objects and the development of collaborative design; (4) Level three, characterized by the integration and interoperability of data via the Internet, with an emphasis on the management of the construction life cycle.

While in the international scenario there is research and development of projects at levels two and three (SUCCAR, 2009; SACKS, 2010; LEE, 2012; ZHANG, 2013), in the Brazilian scenario, despite research already pointing to investigation at levels one and two (COELHO and NOVAES, 2008; MONTEIRO, 2011; ANDRADE, 2012; SANTOS and BARISON, 2013), the maturity of the use of computational tools in Civil Engineering projects is mostly at level zero.

Finally, once BIM has been implemented as a working method, many authors point out that complete adoption does not occur immediately, but over a sequence of stages, until it is fully understood (TOBIN, 2008) (SUCCAR, 2009) (JERNIGAN, 2007). According to RUSCHEL et al., 2011, TOBIN, 2008 and SUCCAR, 2009, the stages of BIM implementation are classified into three:

- INTERNSHIP 1. With a focus on parametric modeling, technology is only used as a specific tool in the graphic representation disciplines, as the work process is still individualized, without the involvement and collaboration of other disciplines in the course. It is usually restricted to a specific phase of the process (design, construction or operation).

- STAGE 2. With a focus on collaboration, the technology has a multidisciplinary nature between one or two phases of the design process, involving up to two different themes, such as architecture and structure, or cost management. The development of the work relies on the interoperability of the design information, making it possible to manipulate a single three-dimensional model between all the teams involved in the work. The process is also interactive and still asynchronous, but with improved interoperability between the teams involved.

- STAGE 3: Focusing on the shared and collaborative creation of the building model synchronously through network integration, the development of the work includes complex analyses already in the early stages of conception, such as the analysis of design conflicts, as the possibilities for interoperability are expanded through open protocols and virtual work environments, so that all the agents involved in the project can contribute collectively within the specificities of their disciplines.

3. BIM IN UNDERGRADUATE TEACHING

With the constant growth in the adoption of BIM by companies in the Architecture, Engineering and Construction (AEC) market, Architecture and Civil Engineering schools have sought to implement subjects in undergraduate courses to expose students to the challenges of developing parametric and collaborative design (HOLLAND, 2010). According to BARISON AND SANTOS (2011), BIM teaching began to be introduced internationally in AEC courses from 2003 onwards, but this practice intensified between 2006 and 2009, in relation to strategies and approaches linked to the level of competence that the student must achieve in relation to the activity that will be carried out in professional practice.

According to Resolution CES/CNE 11 (MEC, 2002), a reference for the preparation of curricula for undergraduate engineering courses, including Civil Engineering, the organization of the curriculum is divided into three levels: (i) the basic content core, which accounts for around 30% of the minimum workload; (ii) the professional content core, which accounts for around 15% of the minimum workload; and (iii) the specific content core, made up of extension activities and in-depth study of the content in the professional content core, which accounts for the rest of the total workload. As it is a set of subjects and/or activities proposed exclusively by the HEI, it covers the scientific, technological and instrumental knowledge needed to define the types of engineering and should guarantee the development of the competences and skills established in these guidelines. Although BIM-related content is not provided for in this resolution, some Civil Engineering courses have approached the subject through research at undergraduate and postgraduate level, as well as experiences of its use in project development disciplines.

However, academic experiences dealing with BIM are new and based on pedagogies that have not yet been consolidated. According to SABONGI (2009), only 9% of North American construction schools teach BIM in their undergraduate courses, the main obstacles being: a lack of time or resources to redesign curricula, as well as a lack of university structure and prepared teachers with specific materials related to teaching the technology. RUSCHEL et al (2013), based on various didactic experiences researched, concluded that, on the national scene, BIM has been implemented gradually and with little effectiveness in Architecture, Engineering and Construction courses, and the main associated problems were the same as those highlighted by SABONGI (2009). It was noted that the international experience is much more mature than that found in Brazil, with BIM being addressed at various stages of the professional's training during the undergraduate course, which is also justified by a market demand in these countries, since the implementation of BIM by international companies has been happening more effectively and quickly.

Therefore, in order to develop a teaching strategy, universities are seeking, as the use of BIM technology spreads, some way of inserting it into the academic context, seeking to understand the role of BIM and how this system has influenced the way of designing; that is, not limiting its use as just a "software package", but also as an exercise in collaboration, sustainability and resource management. (ROMCY et al, 2013 apud KENSEK, 2012).

4. METHODOLOGY

This research was carried out in three stages. In the first stage, a bibliographical survey was carried out on: (a) the conceptualization of BIM, as well as its characteristics and advantages; (b) the evolution of graphical representation technologies in a historical context; (c) the approach to the BIM platform in teaching and in the job market, with a greater focus on education, both nationally and internationally; (d) research in the area of Civil Engineering teaching integrated with BIM; (e) the pedagogical projects of Civil Engineering degree courses at Brazilian universities. In the second stage, data was collected by means of a questionnaire on the academic community's knowledge of BIM. Finally, a qualitative analysis was made of the answers obtained by applying the questionnaires, considering the feasibility of inserting BIM concepts into Design and Project disciplines in an integrated and multidisciplinary way.

The questionnaire used to collect the data for this research initially includes a series of questions designed to get to know the respondent's profile. This stage of the questionnaire included questions about the university to which the respondent is affiliated, the area in which he or she works, the length of his or her training and the length of his or her teaching experience. The questionnaire then asked whether the respondent used BIM technology in their teaching activity. If the answer was yes, a series of questions would be asked about the length of time they had used the BIM tool, their level of experience, their stage of application in the use of modeling in their teaching activities and the changes that were most noticeable after the insertion of BIM in their teaching activity; if not, they would go directly to more general questions on the subject, taking into account their academic experience and knowledge of the subject. Questions about the possible obstacles to implementing technology in education formed part of this stage of the questionnaire. Finally, we asked at what point in the Civil Engineering degree course the lecturer thought the possible inclusion of BIM was most coherent and what aspects of the training of AEC professionals the lecturer thought were relevant in the context of BIM.

This questionnaire was sent via the web to approximately 400 professors of Civil Engineering courses at 42 public universities and 14 private universities in the country, using the e-mail addresses provided on the departments' websites. The computer platform used to send the questionnaire was Google Docs®, as it offers the option of sending it via e-mail or accessing it via a link. It is a free, easy and fast to use platform, whose data is stored in a spreadsheet, automatically generating a summary containing the absolute values, percentages and quantitative graphs of the answers. This questionnaire was available to answer for three months.

5. Results

The results of the questionnaire survey are based on 48 complete answers, which represent 12% of the sample of 400 teachers who were sent the questionnaires. Of the respondents, the majority were from public universities (89.6%), which are places where there is a greater concentration of research and studies in a wide range of areas.

As for the profile of the respondents, 81% answered that they had more than 15 years of undergraduate teaching experience, with no responses indicating less than 5 years of teaching experience. In terms of total teaching experience, 62% have more than 15 years of teaching experience, compared to 17% with between 1 and 5 years of experience in the academic field. By relating these two issues, we can see a common profile in universities: the majority of teachers have a relatively long period of training, as well as teaching experience. This is due to the fact that, in order to become a lecturer in universities, people have to go through the postgraduate levels (master's and doctorate), which take around 6 years to complete. It's rare to find teachers who are very young or who have only been trained for a short time, even in private universities.

As for the area in which they work, the majority are in the field of Civil Construction (35%), followed by Hydraulics, Sanitation and the Environment (19%); Geotechnics (6%); Transportation (10%); Structures (25%) and Architecture and Urban Planning (4%). At first glance, the low percentage of professors working in the area of Architecture and Urbanism might seem to be a misleading figure, since it is the field of activity that is currently most likely to implement BIM compared to the other areas. However, the explanation for this situation is due to the fact that the number of teachers working in this field is low in undergraduate Civil Engineering courses, justified by the reduced number of hours of subjects related to this field, in contrast to the area of Civil Construction, whose subjects form the core of the course.

Regarding the use of BIM technology in teaching, 73% said they did not use it. Of the 27% who said yes, 83% had been using it for less than 5 years. In other words, we can see that BIM technology is still not very widespread in academia and, as a result, its use is still very restricted. This is also due to the fact that it is a recent technology, about which knowledge is still in its infancy.

In order to relate the experience of using the BIM tool in teaching activities to the various academic levels, we asked about activities in the undergraduate and postgraduate areas, subdividing them into teaching and research activities, as well as extension. In terms of teaching at undergraduate level, the

BIM tool is still little covered in most activities, as only 20% of the responses indicate activities with practical and theoretical content on the subject, and 24% do not cover BIM in their area of activity. In research activities at undergraduate level, 14% have the BIM tool as the main theme of the activity. Analyzing postgraduate studies, the same scenario can be observed as in undergraduate studies, as the majority of respondents indicate that they do not address BIM technology in their academic activity. The use of BIM is even more restricted in extension activities, where only 5% use the tool as the main theme of the activity.

For those who approach BIM in their academic activities, according to the stages defined by RUSCHEL *et al* (2011), TOBIN (2008) and SUCCAR (2009), the results obtained suggest that the main focus of the BIM approach is in Stage 1 (58%), characterized by parametric modeling, followed by activities within Stage 2 (26%), focusing on interoperability and asynchronous collaboration, and finally Stage 3 (16%), with synchronous integrated practice.

The questionnaire listed some of the possible beneficial changes that the insertion of BIM would bring with its use, so that the teachers could analyze them and rate them on a scale of 1 to 5, where 1 represents no change and 5 major changes, according to their experience. Thus, 63% were able to see advances in anticipating design problems and making them easier to visualize; 57% said that there had been major improvements in the richness of design details; 37% observed low to medium impact changes in the sense of improved compatibility between designs and 47% said major improvements in the reduction of graphical representation errors. In this way, we can see that, in all the points analyzed, improvements in project development were indicated with the use of the tool. In addition, a space was opened up for teachers to add other changes that they considered important, and among the changes listed, the most cited were improved teamwork and an integrated view of the project's life cycle using BIM technology.

Specifically regarding the implementation of BIM technology in the academic area, some aspects are considered influential, and these were classified by the respondents on a scale of 1 to 5, where 1 represents not very influential and 5 very influential. The most influential aspect for the implementation of BIM technology was the availability of BIM teaching infrastructure in universities, such as laboratories, hardware and software, as 94% of the responses indicate that this is a highly important issue in the implementation of BIM in the academic area. The next most influential aspects were: greater integration between course subjects (81%), compatibility with CAD tools that have already been consolidated (79%), access to teaching materials (73%) and resistance to adopting BIM, as it is a technology that has not yet been consolidated in academia (69%). The respondents were also able to add other influential aspects to the implementation of BIM in universities, including the need to engage the teaching staff in changing methodology and specific standards on the subject.

Finally, in relation to the best time in the Civil Engineering course curriculum for BIM content to be adopted, based on Resolution CES/CNE 11 (MEC, 2002), the following answers were obtained: 48% believe that BIM should be implemented in the specific content core; 41% in the professionalizing content core and 10% in the basic content core. In addition to this question, some lecturers stressed that BIM content should be included at more than one point in the Civil Engineering course, while others believe that it should be covered in an informative way, playing a supporting role in the Civil Engineering teaching process.

The graphs containing the results of the questionnaire responses can be seen in Annex 1.

6. Conclusion

According to the data collected by applying the questionnaires, it is possible to observe that knowledge of BIM technology is still restricted, even in the academic area, where there is a greater chance of approaching new technologies and new horizons of knowledge. This is due to the fact that the BIM paradigm is recent and not yet consolidated, and is being introduced, both in education and in the job market, at a slow and conservative pace in Civil Engineering (RUSCHEL, R. *et al*, 2013). This is a reality that is directly associated with the academic community's resistance to change and lack of knowledge of how to use BIM technology.

In order to implement BIM teaching in the academic environment, the infrastructure of universities is a crucial point, since it was the aspect considered most relevant by the teachers who answered the questionnaire. For this reason, one of the major obstacles to the introduction of BIM in Civil Engineering courses is the lack of structure in terms of physical training, such as laboratories, hardware and software, and the professional training of teachers and technicians.

Given this scenario, when thinking about when BIM concepts should be included in Civil Engineering courses, first of all, according to BARISON and SANTOS (2011), we need to know what levels of BIM knowledge the student should have, and how much time will be allocated in the curriculum to teaching this subject. Once this aspect has been defined, one strategy pointed out in the results is to include BIM in the three levels of the course curriculum, with each approach being compatible with the student's level of education and training needs.

In this way, BIM, at an introductory level, should be included in the basic cycle of the Civil Engineering course curriculum, associated with Graphic Expression subjects such as Computer Aided Design and Technical Drawing, including an initial approach to the concept of BIM and parametric modeling, equivalent to Stage 1. BIM can also be introduced in the vocational content core, since this stage of the course requires the student to be more knowledgeable about materials and construction methods, investing in knowledge about modeling, teamwork, conflict analysis and integration between disciplines, in an approach equivalent to Stage 2. Finally, at the end of the degree course, it would be important to take an advanced approach to BIM, equivalent to Stage 3, exploring the principles of interoperability, management, project conflicts and simulations in 4D (which considers, in addition to project dimensions, the "time" variable in modeling) and 5D (which considers the "cost" variable in the project, in addition to dimensions and time). Therefore, as this is a gradual and evolutionary process, each level should be introduced according to the availability of resources at the universities, as well as the faculty's knowledge of the subject.

Finally, this work, together with the others analyzed in the literature review, supports the teaching of BIM as an essential subject for training professionals in Civil Engineering. It is believed that knowledge of this technology will soon be a decisive factor in hiring by AEC companies, so it is important that universities are aligned with the demand for knowledge required by the job market and, in this way, prepare their teaching staff, students and their physical and curricular structure to meet these objectives.

Thanks

The authors would like to thank the CNPq - National Council for Scientific and Technological Development, which supported this research with a grant from the Unified Program for Scientific and Technological Initiation.

REFERENCES

ANDRADE, Max Lira Veras Xavier de. Performative project in recent architectural practice: Conceptual framework. Campinas, SP: [s.n.], 2012. PhD Thesis - State University of Campinas, Faculty of Civil Engineering, Architecture and Urbanism.

BARISON, M. B.; SANTOS, E. T. Current Trends in BIM Teaching. V TIC - Meeting of Information and Communication Technology in Construction. Salvador, 2011.

CEROVSEK, Tomo. A review and outlook for a 'Building Information Model' (BIM): A multi-standpoint framework for technological development. Advanced Engineering Informatics 25 (2011) 224-244

COELHO, Sérgio Salles; NOVAES, Celso Carlos, Building Information Modeling (BIM) and Collaborative environments for project management in civil construction. (2008) Available at: <www2.pelotas.ifsul.edu.br/gpacc/BIM/referencias/

COELHO_2008.pdf> Accessed May 2013.

EASTMAN, Chuck et. al. (2011) BIM Handbook: A Guide to Building Information Modeling for Owners, Managers, Designers, Engineers and Contractors, John Wiley & Sons Inc, New Jersey, 2nd edition.

HABITARE. Innovation, Quality & Productivity Management and Knowledge Dissemination in Housing Construction - Vol. 2. 2003a. Accessed on: 20 Oct 2008. Available at: http://www.habitare.org.br/publicacao_coletanea2.aspx

. Standardization and Certification in Housing Construction - Vol.3. 2003b. Accessed on: 20 Oct 2008. Available at: http://www.habitare.org.br/publicacao_coletanea3.aspx

HOLLAND, R.; MESSNER, J.; PARFITT, K.; POERSCHKE, U.; PIHLAK, M.; SOLNOSKY, R. Integrated Design Courses Using BIM as the Technology Platform. Available at: <http://php.scripts.psu.edu/users/r/l/rls5008/CV/Linked%20Papers/Holland%20et%20al%202010%20 Integrated%20Design%20Courses%20Using%20BIM%20as%20the%20Technology%20Platform.pdf > Accessed: May 10, 2014.

JERNIGAN, F. Big BIM little BIM: the practical approach to Building Information Modeling integrated practice done the right way! Salisbury: 4 Site Press, 2007. 323 p.

KENSEK, K. M. Advancing BIM in Academia: Explorations in Curricular Integration. Computational Design Methods and Technologies: Applications in CAD, CAM and CAE Education. IGI Global, 2012.

LEE, Seung-Il et. al. Efficiency analysis of Set-based Design with structural building information modeling (S-BIM) on high-rise building structures. Automation in Construction 23 (2012)20-32

MEC - Ministry of Education. Resoluçao CNE/CES n^0 11, de 11 de março de 2002 " Institui Diretrizes Curriculares Nacionais do Curso de Graduaçao em Engenharia. Brasilia, 2002.

MONTEIRO, Ari. Design for the production of vertical masonry walls using a CAD-BIM tool. 2011. Master's degree dissertation. Polytechnic School of the University of Sao Paulo. Department of Civil Construction Engineering.

PORWAL, Atul, HEWAGE, Kasun N. Building Information Modeling (BIM) partnering framework for public construction projects. Automation in Construction 31 (2013) 204-214

ROMCY, N. M. S.; CARDOSO, D. R.; MIRANDA, N. M. BIM and Teaching: Academic Experience at the Federal University of Ceara. VI TIC - Meeting of Information and Communication Technology in Construction. Campinas, 2013.

RUSCHEL, R. C.; ANDRADE, M. L. V. X.; SALES, A. A.; MORAIS, M. The Teaching of BIM: Examples of Implementation in Architecture Engineering Courses. V TIC - Meeting of Information and Communication Technology in Construction. Salvador, 2011.

SABINO, W. CAD-BIM-GIS: Integrating Solutions. 2008. Accessed on: 20 Oct 2008.

Available at: http://pasta.ebah.com.br/download/cad-bim-gis-integracao-de-solucoes-wolmar-sabino-ppt-7866

SABONGI, F. J. The Integration of BIM in the Undergraduate Curriculum: an analysis of undergraduate courses. In: 45th Annual Conference of Associated Schools of Construction, 2009. Available at: <http://ascpro0.ascweb.org/archives/cd/2009/paper/CEUE90002009.pdf> Accessed on: May 28, 2014.

SACKS, Rafael et. al. Requirements for building information modeling based lean production management systems for construction.Automation in Construction 19 (2010) 641-655

SANTOS, Eduardo Toledo; BARISON, Maria Bernardete. Bim and universities. Construçao e Mercado magazine. Available at: <http://revista.construcaomercado.com.br/negocios- incorporacao-construcao/115/o-desafio-para-as-universidades-formacao-de-recursos- humanos-208417-1.asp> Accessed May 2013.

SMITH, D. K. and TARDIF, M. (2009) Building Information Modeling: A Strategic Implementation Guide for Architects, Engineers, Constructors, and Real Estate Asset Managers, John Wiley & Sons Inc.

SUCCAR, Bilal. Building information modelling framework: A research and delivery foundation for industry stakeholders. Automation in Construction 18 (2009) 357-375

TOBIN, J. Building Information Modeling for tertiary construction education in Hong Kong. Journal of Information Technology in Construction. Sept. 2011. Available at: <http://www.itcon.org/cgi-bin/works/Show72011_27> Accessed on: May 10, 2014.

ZHANG, Sijie et. al. Building Information Modeling (BIM) and Safety: Automatic Safety Checking of Construction Models and Schedules. Automation in Construction 29 (2013) 183195

BIM ISSUES IN THE CIVIL ENGINEERING UNDERGRADUATE COURSE

Abstract: This article is about the actual skills of undergraduate students of Civil Engineering Course related to BIM issues. The research goal nus to find pedagogical strategies to teach BIM concept in undergraduate course, considering the Brazilian normative resolutions (MEC, 2002). Through bibliography research and questionnaire evaluation, answered by professors of Brazilian universities, it is possible to suggest a strategy that inserts BIM topics in three different stages of undergraduate course, according to the level skills of the student and his formation needs. Finally, this work, agrees with the bibliography review, supporting that the BIM teaching as an essential issue for professional qualifications of Civil Engineering undergraduate students.

Key-words: BIM; BuildingInformationModelling; CivilEngineering; Undergraduate course

ANNEX 1 - GRAPHS OF THE QUESTIONNAIRE RESULTS

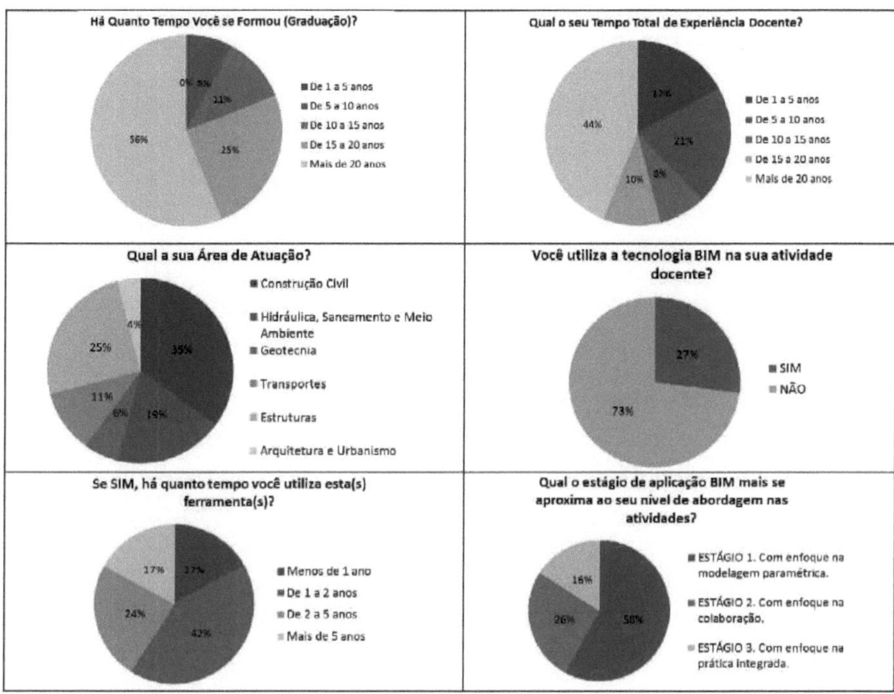

ANNEX 1 - Graphs of the questionnaire results (continued)

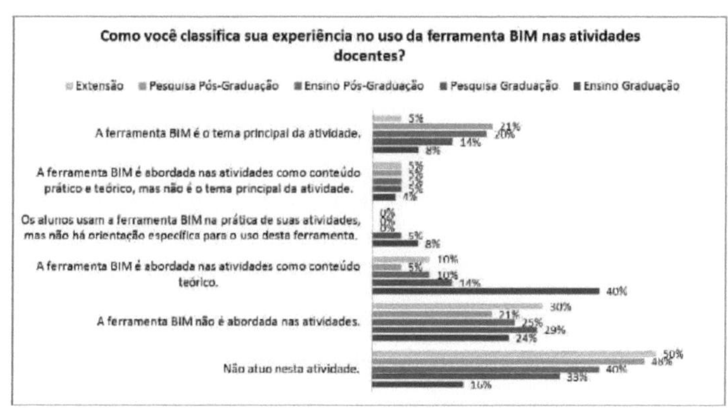

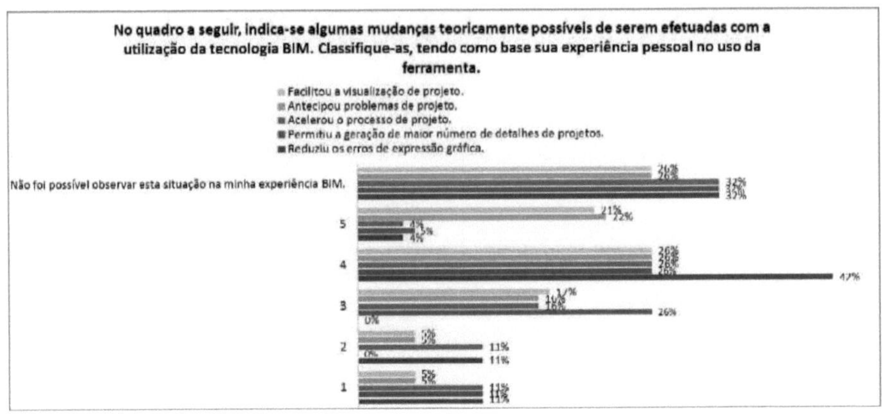
No quadro a seguir, indica-se algumas mudanças teoricamente possíveis de serem efetuadas com a utilização da tecnologia BIM. Classifique-as, tendo como base sua experiência pessoal no uso da ferramenta.
Facilitou a visualização de projeto.
Antecipou problemas de projeto.
Acelerou o processo de projeto.
Permitiu a geração de maior número de detalhes de projetos.
Reduziu os erros de expressão gráfica.
Não foi possível observar esta situação na minha experiência BIM.
5
4
3
2
1

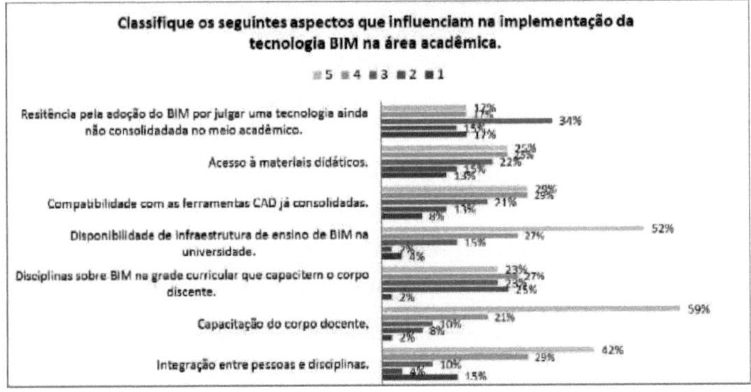
Classifique os seguintes aspectos que influenciam na implementação da tecnologia BIM na área acadêmica.
5 4 3 2 1
Resitência pela adoção do BIM por julgar uma tecnologia ainda não consolidadada no meio acadêmico.
Acesso à materiais didáticos.
Compatibilidade com as ferramentas CAD já consolidadas.
Disponibilidade de infraestrutura de ensino de BIM na universidade.
Disciplinas sobre BIM na grade curricular que capacitem o corpo discente.
Capacitação do corpo docente.
Integração entre pessoas e disciplinas.

Printed by Books on Demand GmbH, Norderstedt / Germany